YOUR KNOWLEDGE HAS VALUE

- We will publish your bachelor's and
  master's thesis, essays and papers

- Your own eBook and book -
  sold worldwide in all relevant shops

- Earn money with each sale

Upload your text at www.GRIN.com
and publish for free

John Tarilanyo Afa

The Lightning Performance of High Voltage Transmission Line

GRIN Verlag

Bibliografische Information der Deutschen Nationalbibliothek:

Die Deutsche Bibliothek verzeichnet diese Publikation in der Deutschen National-
bibliografie; detaillierte bibliografische Daten sind im Internet über http://dnb.d-
nb.de/ abrufbar.

Imprint:

Copyright © 2011 GRIN Verlag GmbH
Druck und Bindung: Books on Demand GmbH, Norderstedt Germany
ISBN: 978-3-656-41353-0

JOHN TARILANYO AFA

CURRICULUM DESIGN

TECHNICAL REPORT

LPH 657:

STUDY ON THE LIGHTNING PERFORMANCE OF HIGH VOLTAGE TRANSMISSION LINE

ATLANTIC INTERNATIONAL UNIVERSITY

HONOLULU, HAWAII

2011

TABLE OF CONTENT

LIGHTNING PERFORMANCE OF H.V TRANSMISSION LINE

1.0   INTRODUCTION

Lightning is a major source of danger to H.V transmission lines resulting in serious overvoltage which may cause flashover, puncture and sometimes loss of transmission line up to few hours or complete destruction of lines.

The study is focused on Bayelsa State whose coastal areas occupy about eighty percent of the state. Due to its geographical location (Mangrove swamp forest) especially the coastal areas, these are areas of high thunderstorm and lightning days per year. This will pose significant influence on transmission line performance.

From the present load with the proposed projects when completed may demand bulk power supply. The need of using high voltage transmission lines is there. Presently, it is expected that 132kv line may run from Yenagoa to Brass through Ogbia. It is therefore necessary to study the line performance in these areas of high Isokeraunic level. Lightning faults may be single or multi-phase and their elimination causes reclosing cycles, voltage dips and outages. Therefore the outage rate of a line and the quality of the delivered voltage depends on the lightning performance of the line. No transmission line design can be considered lightning proof, nor designers aimed at this goal. An acceptable design is to allow a certain number of outage per 100km of line or other line durations (Razevig 2003, Uglesic 2009). The probability of an outage depends on many factors which are statistical in nature that a worst case design is neither practical nor economical.

1.1   PREDISCHARGE CURRENT MECHANISM

When considering lightning stroke mechanism it is necessary to look at the phenomenon called the pre-discharge current. As the transmission voltage increases it also necessitates a higher tower, the width and breadth also increase. During the discharge of lightning the ground plane will be elevated to the metallic parts of the transmission line at the tower and at the same time the infinity flat ground surface being replaced by metallic parts of smaller radius of curvature yielding very high stress concentration which the lightning stroke looks from above. At the high

electric fields present, a corona discharge develops at the tips of the tower and this also propagates upward to meet the down coming leader and its corona envelop. They meet at a height of about 60 to 100 meters above earth.

Back flashover mechanism became important in which the tower top potential at the location of outer phases increased to such a high value as to cause an insulation flashover due to high pre-discharge currents induced by high fields at the pointed tip of the cross arm (Mushin-Tun,2008,Pinto et al 2006). Due to travelling wave effect inside the long frame work of the tower itself, the tower top potential is high and it takes time for all reflections to die down sufficiently to consider the tower as acquiring ground potential even though it is grounded well at the tower footing.

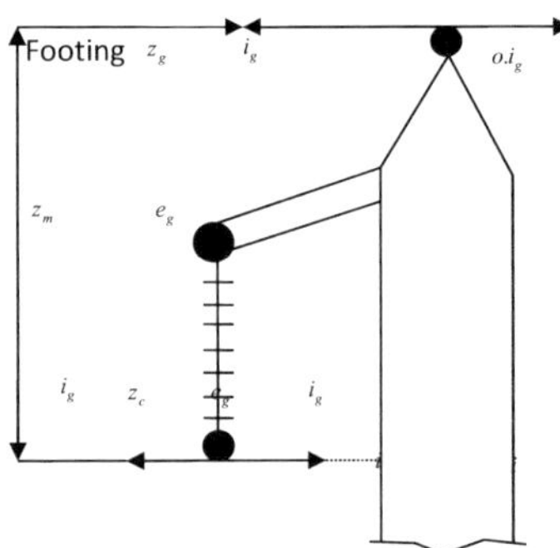

Fig. 1: Voltage Stress on insulators

When a stroke contacts a tower, the voltage stress experienced by the insulator can be found.

$$e_g \;=\; Z_g\, i_g \;+\; Z_m\, i_c \quad and \quad e_c = Z_m\, i_g \;+\; Z_c\, i_c$$

$$e_c \;=\; k_f\, e_g \;+\; \left(z_c - k_f\, Z_m\right) i_c$$

Where k_f is coupling factor between ground wire and the live conductor.

If the surge impedances are equal $\left(z_g = z_c\right)$, then

$$E_c = k_f E_g + \left(i - k_f^2\right)z_c\, i_c \quad - \quad - \quad - \quad - \quad - \quad - \quad - \quad - \tag{1}$$

The voltage across the insulation string is

$$E_2 = E_g - E_c = \left(1 - k_f\right)E_g - \left(1 - k_f\, z_m / z_c\right)z_c\, i_c \quad - \quad - \quad - \quad - \tag{2}$$

The second term is caused by the presence of pre-discharge current contributed by the line conductor. If the pre-discharge current can artificially increased, the insulator voltage is lowered.

It was observed (Begamudre 2008, Ekonomouet al,2003) that $i_c = 0.5 i_{pd}$

Where i_c is the pre-discharge current flowing in the conductor

The second term gives an effective impedance of

$$Z_e = \tfrac{1}{2} Z_c \left(1 - K_f\, Z_m / Z_c\right) \quad - \quad - \quad - \quad - \quad - \quad - \quad - \tag{3}$$

The major factors affecting the lightning performance of a transmission lines can be summarized as follows:

(i)   Isokeraunic level

(ii)   Magnitude and wave shape of current

(iii)   Tower height

(iv)   Resistance of tower and its footing

(v)   Number and location of over head ground wire (shielding angle of conductors)

(vi)   Span length

(vii)   Mid-span clearance between conductors and overhead ground wire.

(viii) Number of insulator units.

Since the basic input such as the lightning frequency, lightning current magnitude, wave front time, and incident rate are random variables, the predictions of lightning performance of a line is a probability problem.

Therefore, various probabilistic methods of computing the lightning flashover performance of lines have been developed.

2.0 GENERAL ANALYSIS AND CALCULATION PROCEDURES

From the workdone by the IEEE working group based on the work of Anderson (Ekonomou et al 2006,Karampelas et al 2010), the calculation was in line with the flash programme. The first step is to characterize the lightning activities in the region crossed by the line, that is, to know the ground flash density N_g. Since local measurements were not there. N_g was calculated from the thunderstorm days (T_d) or the thunderstorm hours (T_h)

$$N_g = 0.04T_d \qquad\qquad (4)$$

The number of flashes to earth that intercepted a transmission line

N_L, that is number of flashes per 100km lightning days per year is

$$N_L = N_g / _{10} \left(4h^{1.09} + g\right) \qquad\qquad (5)$$

where h, is the tower height in meters g is the horizontal spacing in meters between the shielding wires.

The impulse ground resistance is less than the measured or calculated resistance because significant ground current cause voltage gradients sufficient to breakdown the soil around the ground rod. The decreased tower footing resistance R_t when lightning current amplitude exceeds a critical value I_s is given by

$$R_t = \frac{R_o}{\sqrt{1 + \frac{I}{I_s}}} \qquad\qquad (6)$$

But $I_s = \frac{1}{2n}\frac{(E_o \rho_o)}{R_o^2}$

where R_o is the tower footing resistance (lump) at low current

I_s is the limiting current to initiate sufficient soil ionization

I is the stroke current through the resistance

ρ is the soil resistivity (ohm-m)

E_o is the soil ionization gradient (app. 300kv/m).

The simplified expression for insulator withstand capability can be calculated as

$$V_f = \left(K_1 + \frac{K_2}{t^{0.25}} \right) L \qquad (7)$$

Where V_f is a flashover voltage and K_1 is 400, K_2 700 and L is the length of insulator and t is elapse time after lightning stroke in uS.

The termination point of a lightning stroke to a transmission line can be either a shielding wire, phase conductor, tower or even ground. The electrogeometric model using the concept of stricking distance could generally be given as S by the formular Gonos et al 2003, Martinez and Castro-Aranda, 2007, He,et al 2005)

$$S = A.I^b \qquad (8)$$

Where A and b are constants depending on the point of termination, and I is the prospective stoke current in KA. From the table 1 the constant for A and b are given by different sources.

Source	Stricking Distance to ground		Stricking distance to phase conductor and shield wire	
	A	B	A	B
IEEE WG	6.4	0.65	0.8	0.65
Amstrong and white head	60	0.8	6.7	0.80
Brown and whitehead	6.4	0.75	7.1	0.75
Love	10	0.65	10.0	0.65
Rizk	-	-	$1.5h^{0.5}$	0.69

Using the improved model proposed by Erikson (Ab Kadir 2009, Ekonomou et al 2003), considering the attractive (critical) radius around the transmission lines

$$S = 0.67 H^{0.6} I^{0.74} \qquad (9)$$

H is the structure height in meters and I is the prospective stroke current in KA.

The shielding failure flashover rate N_{sf} can be estimated by

$$N_{sf} \;=\; N_L \cdot \int_{I\min}^{I\max} D_s \; f(I)dI \qquad\qquad (10)$$

where I_{max} is the maximum lightning current in KA, I_{min} is the minimum current equal to $2U_a / Z_s \; (surge) U_a$ is the insulation level of the transmission line in KV, $f(I)$ the current density probability and S is the shielding failure exposure distance and Z_s is the conductor surge impedance.

Back flashover failure rate BF is estimated for transmission lines as

$$N_{BF} = N_L \cdot \int_0^\infty P(\delta) \qquad\qquad (11)$$

where $P(\delta)$ is the probability distribution function of δ, δ is an auxiliary variable given in KV.

$$\delta \;=\; R \cdot \frac{I\max}{2} - 0.85.V_a + L \frac{dL}{dt} \qquad\qquad (12)$$

where R is the footing resistance, L the total inductance of Tower and grounding system.

The total outage rate is the sum of failure shielding failure rate N_{sf} and the back flashover failure rate N_{Bf}

$$N_T = N_{SF} + N_{BF} \qquad\qquad (13)$$

3.1   MATERIALS AND METHODS

The technical characteristics including line configuration and grounding impedance were studied.

The resistivity test was carried out in six locations of the line route.

The tower structure is shown in fig. 2 with some tower parameters. The tower carries three phase single circuit, with double earth wire. Other details of line and structure are shown in table 2.

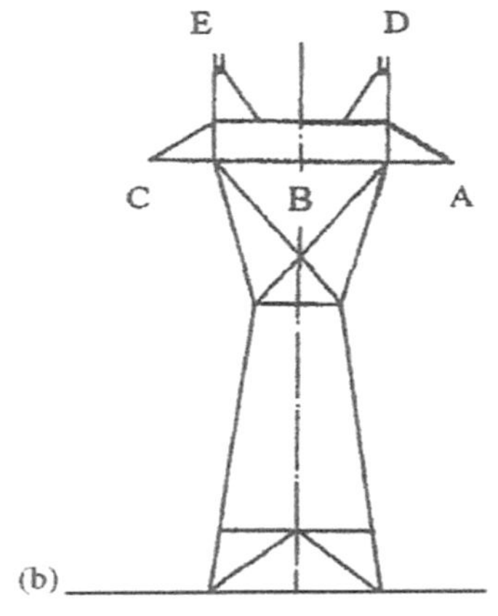

E = D = 4.8m x 23m

A = C = 7m x 21.74m

B = 0.0m x 21.74

ED = (distance) = 8.4m

Fig. 2: Pole Type

Table 2 Line and Tower Parameters

Line length	116.4km
Conductor Data	1 x 500mm^2
Insulator Data	10 disc x 146mm
Line Sag	8.2 meters
Tower Height	23 meters
Phase wire height	21 meters
Ground Resistance in Ω	3-6Ω
Soil resistivity (measured)	50-120 Ωm
Normal Spam length	340m
BIL	550KV
No. of Shielding Wire	2

3.2   METHODOLOGY:

To achieve the objective of the study, the following parameters were known or calculated.

- ❖   No of shield wire and separation among shield wires
- ❖   Average height of shield wire
- ❖   Length of line or section detail
- ❖   Days of thunderstorm per year
- ❖   Probability of exceeding stroke current
- ❖   Rate of line trip out per year

The first requirement is to know the ground flash density which can be calculated either using the keraunic level or by estimation if the optical transient density is known for the specific area. More specifically the keraunic level can be assessed either from Isokeraunic Maps or from daily weather records being obtained from ground based observations.

This line runs across the coastal areas with some river crossing. The kerounic level in this area is high but an average of 120 thunderstorm per day was used in the calculations.

4.0   RESULTS

The soil resistance and resistivity values are shown in table 3.

	Yenagoa		Ogbia		Brass	
Season	Resistance Ω	Resistivity Ω - m	Resistances Ω	Resistivity Ω - m	Resistance Ω	Resistivity Ω - m
Dry	6.5	120	5.0	90	3.0	50
Wet	4.2	80	4	60	2.5	40

The trip out of lines was calculated using a matlab programme as shown in the flow chart (fig. 4)

Line Details Data

NT	-	Number of Towers
LL	-	Line Length
CD	-	Conductor Data
ID	-	Insulator Data
LS	-	Line Sag (max)
NG	-	Average Ground Flash Density (optical Transient Density
	OR	
T	-	Keramic Level
NTr	-	No of Tripping
GR	-	Ground Resistance
SR	-	Soil Resistance
	OR	
TR	-	Tower Resistance
B	-	Number of Shield Wires
h	-	Height of Tower (in meters)
P_f	-	Probability Density function

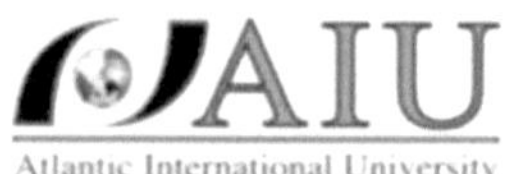

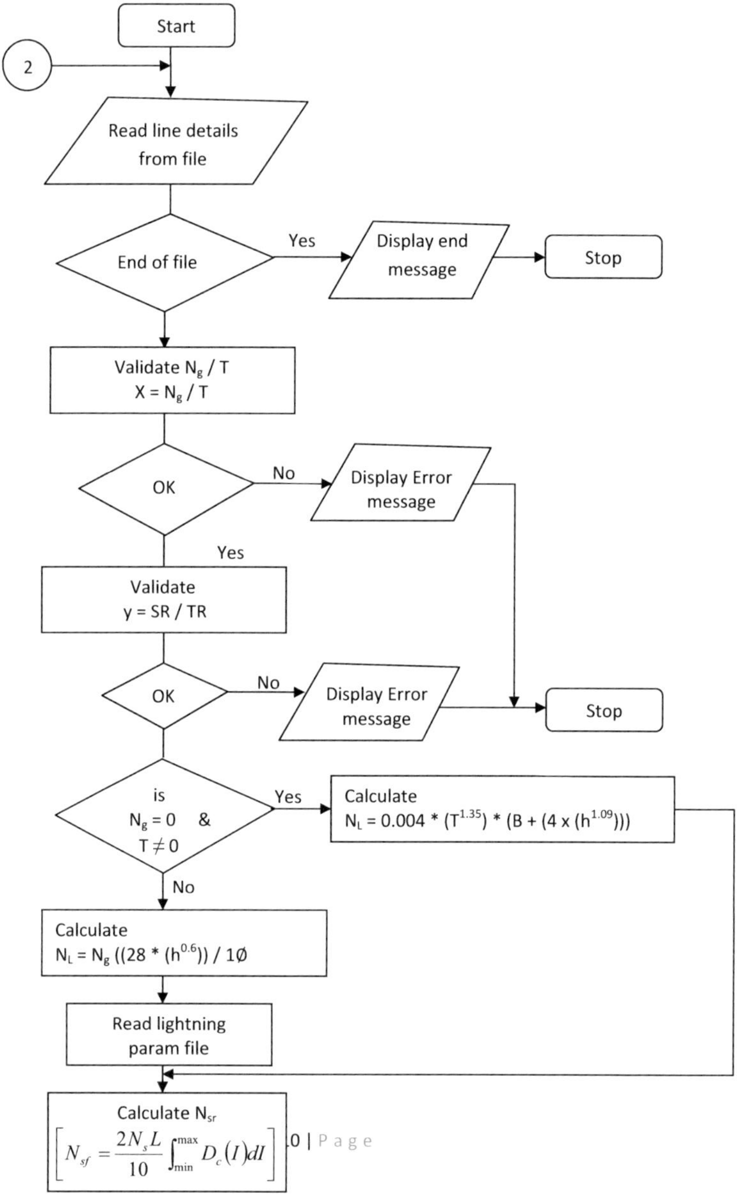
Start
2
Read line details from file
End of file
Yes
Display end message
Stop
Validate N_g / T
X = N_g / T
OK
No
Display Error message
Yes
Validate
y = SR / TR
OK
No
Display Error message
Stop
is
$N_g = 0$ &
$T \neq 0$
Yes
Calculate
$N_L = 0.004 * (T^{1.35}) * (B + (4 x (h^{1.09})))$
No
Calculate
$N_L = N_g ((28 * (h^{0.6})) / 1Ø$
Read lightning param file
Calculate N_{sr}
$\left[N_{sf} = \frac{2N_s L}{10} \int_{min}^{max} D_c(I) dI\right]$

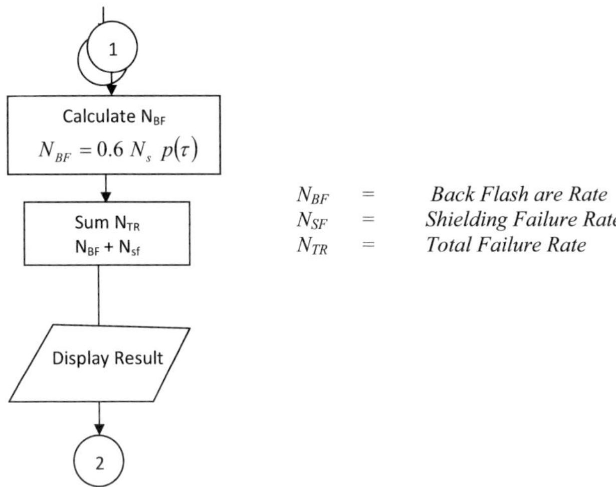

Fig. 4: Flow Chart for Total Line trip out

The number of lightning flashes to the line per 100km per year = 30 Strokes

The back flashover N_{BF} = 1.25 per 100km / year

The shielding failures N_{SF} = 0.9 per 100km / year

The Rainy condition trip out = 4.82 per 100km / year

5.0   DISCUSSION:

From the results the trip out rate is small because of the low tower footing resistance. The lower the tower footing resistance the more negative reflections produced from the tower base towards the tower top and these hence help to lower the peak voltage at the lower top. Also, the influence of the tower footing resistance on the tower top is determined by its value which is also surge current dependent. The larger the surge current, the smaller the tower footing resistance and as such more negative reflections are produced.

The use of two ground wire was as a result of the high keraunic level. With two ground wire the ground flash density (lightning days / km^2 /year) is high but provide adequate protection against the shielding failure rate.

The longer the span length the more probable the effect of shielding failure due to swing as a result of wind effect therefore the span length was kept at 340 meters.

6.0   RECOMMENDATIONS:

❖   To avoid strokes incidence in phases, install the shield wires in a position that guarantees an effective shielding. That eliminates the trip out by shielding failure. That is, the shielding angle (α) should be calculated that it makes a perfect shield on the line.

❖   The footing resistance should also be kept low (at least 10), to reduce the trip out rate per 100km per year.

❖   Due to the nature of soil regular (ones a year) test should be carried out on the resistivity (resistance) to be sure of low resistivity on the line.

7.0   CONCLUSION:

It shows clearly that the variation of the transmission line insulation level and the tower footing resistance influence significantly the shielding failure and the back flashover failure rate. The total flashover rate or the outage rate is the arithmetic sum of the shielding failure N_{sf} and the back flashover failure rate N_{BF}.

For a more effective protection it is necessary to look into the normal and adverse weather conditions including pollution due to dust, storm, humidity and temperature variations as well as the thunderstorm occurrence. The combined factors may severely affect the transmission lines failure rate estimations.

It is therefore necessary to further investigate the most appropriate selection design parameters in order to reduce the lightning failure rate.

The present line is a proposed line therefore the records will serve as a guide to designers and live construction engineers, therefore if proper selection is made on the shielding angle and proper maintenance of the earthing system are done trip out rate will be minimal in this environment of high keraunic level.

8.0 REFERENCE

1. Abkadir, M.Z.A, Sardi, J., Wan Ahmad, W.F., Hizam, H., Josni, J., 2009, Evaluation of 132kv Transmission Line Performance Via Transient Modelling Approach, European Journal of Scientific Research 29 (4): 533 – 539.

2. Begamudre, R.D., 2008 Extra High Voltage A.C Transmission Engineering. New Age Intrnational Publishers – New Delhi.

3. Ekonomou, L., Gonos, I.F., Stathopulos, I.A., Topalis, F.V., 2003. Lightning Performance evaluation of Hellenic High Voltage Transmission Lines. xiii[th] International Symposium on High Voltage Engineering, Netherland, Smit(ed): 1 – 4.

4. Ekonomou, L., Iracleous, D.P., Gonos, I.F., Stathopulos, I.A., 2005. An optimal Design Method for Improving the Lightning Performance of Overheat High Voltage Transmission Lines ELSEVIER – Electric Power System Research 76: 493 – 499.

5. Gonos, I.F., Ekonomov, L., Topalis, F.V., Stathopulos, I.A., 2003. Probability of Back flashover in Transmission lines due to lightning strokes using monte-carlo Simulation: Internal .J. Elects. Power Energy Sys. 25(2) 107 - 111

6. Gustavo carrosco, H., Alessandro villa, R., 2003. Lightning performance of transmission line LAS CLARITAS – SANTA ELENA UP 230kv. International Conference on Power System Transmission IPST 2003 in New Orleans.

7. Halasa, G., Badran, I., El-Zayyat, H., 2007. Lightning over-voltage on Amman-Agaba 400kv line. American Journal of Applied Science 4 (12): 1075 – 1078.

8. He, J. Tu. Y. Zeng, R., Lee, L.B., Chang, S.H., Guan, Z. 2005: Numerical Analysis for shielding failure of transmission line under lightning stroke: IEEE Trans. Power Deliv. 20(2) 815 – 802.

9. Karampelas, P. Ekonomou, L., Panetsos, S., Chatzarakis G.E., 2010. An Interactive Simulation tool for assessing the lightning performance of Hellemic High Voltage Transmission Lines. ELSEVIER – Applied Soft Computing. 11, 1380 – 1387.

10. Martinez, J.A., Castro-Aranda, P., 2003: Lightning Performance Analysis of Transmission Lines using the EMTP, Power Eng. Soc. Gen. Meet. 1 (2003) 295 – 300.

11. Martinez J.A., Castro-Aranda, P., 2007: Modeling Overhead transmission lines for line arrestor studies: IEEE Power Engineering Society General Meeting.

12. Mowete, A.I., Adelabu, M.A.K., 2009. AN Assessment of Lightning and other Electrostatic Disturbances Along the Coast of South West Nigeria. XIXth International Conference on Electromagnitic Disturbance Bialystok, Poland. 23 – 25.

13. Mushin, Tun (2008). The pollution flashover on high voltage insulations. ELSEVIER – Electric Power System Research 78: 1914 – 1921.

14. Pinto, I.R.C.A., Pinto, O., Naccarato, K.P., 2006. How Ground flash density obtained by lightning location network can be used in lightning protection standards: A Case Study in Brazil. 19th International lightning detection conference Tucson Arizom USA.

15. Razevig, D.V., 2003. High voltage Engineering Khanna Publishers, Delhi.

16. Uglesic, I., Xemand, A., Milardic, V., Milesevic, B., Filipovic-Greic, B., Line. International conference on Power System Transient (IPST 2009) in Kyoto Japan.

17. Yadee, P., Premrudea preechacharn 2007: International Conference on Power System Transients (IPST '07) in Lyon France.